CRANBERRY AND BLUEBERRY CULTURE

WITH INFORMATION RELATING TO GROWING FOR PROFIT

BY

CHARLES S. BECKWITH
&
CHARLES A. DOEHLERT

Copyright © 2013 Read Books Ltd.
This book is copyright and may not be
reproduced or copied in any way without
the express permission of the publisher in writing

British Library Cataloguing-in-Publication Data
A catalogue record for this book is available from the
British Library

CONTENTS

INTRODUCTION TO FRUIT GROWING 1

CRANBERRY CULTURE 5

BLUEBERRY CULTURE 34

Introduction to Fruit Growing

In botany, a fruit is a part of a flowering plant that derives from specific tissues of the flower, one or more ovaries, and in some cases accessory tissues. In common language use though, 'fruit' normally means the fleshy seed-associated structures of a plant that are sweet or sour, and edible in the raw state, such as apples, oranges, grapes, strawberries, bananas, and lemons. Many fruit bearing plants have grown alongside the movements of humans and animals in a symbiotic relationship, as a means for seed dispersal and nutrition respectively. In fact, humans and many animals have become dependent on fruits as a source of food. Fruits account for a substantial fraction of the world's agricultural output, and some (such as the apple and the pomegranate) have acquired extensive cultural and symbolic meanings. Today, most fruit is produced using traditional farming practices, in large orchards or plantations, utilising pesticides and often the employment of hundreds of workers. However, the yield of fruit from organic farming is growing – and, importantly, many individuals are starting to grow their own fruits and vegetables. This historic and incredibly important foodstuff is gradually making a come-back into the individual garden.

The scientific study and cultivation of fruits is called 'pomology', and this branch of methodology divides fruits into groups based on plant morphology and anatomy. Some of these useful subdivisions broadly incorporate 'Pome Fruits', including apples and pears, and 'Stone Fruits' so called because of their characteristic middle, including peaches, almonds, apricots, plums and cherries. Many hundreds of fruits, including fleshy fruits like apple, peach, pear, kiwifruit, watermelon and mango are commercially valuable as human food, eaten both fresh and as jams, marmalade and other preserves, as well as in other recipes. Because fruits have been such a major part of the human diet, different cultures have developed many varying uses for fruits, which often do not revolve around eating. Many dry fruits are used as decorations or in dried flower arrangements, such as lotus, wheat, annual honesty and milkweed, whilst ornamental trees and shrubs are often cultivated for their colourful fruits (including holly, pyracantha, viburnum, skimmia, beautyberry and cotoneaster).

These widespread uses, practical as well as edible, make fruits a perfect thing to grow at home; and dependent on location and climate – they can be very low-maintenance crops. One of the most common fruits found in the British countryside (and towns for that matter) is the blackberry bush, which thrives in most soils – apart from those which are poorly drained or mostly made of dry or sandy soil. Apple trees are, of course, are another classic and whilst they may take several years to grow into a well-established tree, they will grow nicely in most sunny and well composted areas. Growing one's own fresh, juicy tomatoes is one of the great pleasures of summer gardening, and even if the gardener doesn't have room for rows of plants, pots or hanging baskets are a fantastic solution. The types, methods and approaches to growing fruit are myriad, and far too numerous to be discussed in any detail here, but there are always easy ways to get started for the complete novice. We hope that the reader is inspired by this book on fruit and fruit growing – and is encouraged to start, or continue their own cultivations. Good Luck!

CRANBERRY CULTURE

CHARLES S. BECKWITH

A member of the staff of the New Jersey Agricultural Experiment Station for nearly three decades, Charles S. Beckwith was cranberry specialist from 1920 to 1931 and has served as chief in charge of cranberry (and blueberry) research for 13 years. He has been secretary of the American Cranberry Growers' Association since 1923. Trained as an entomologist at Rutgers, he does his research at the substation situated in the heart of the cranberry-growing area, and on the bogs of cooperating growers. Investigations directed by Mr. Beckwith have provided, among other benefits, methods of control for the blunt-nosed leaf hopper, carrier of the false blossom disease, and for other insects; improvements in fertilizer formulas and practice; principles in the use of flooding water on cranberry bogs; and a method of overhead irrigation to prevent frost damage.

THE growing of cranberries is a well-established business in limited sections of the United States and has been carried on for about one hundred years. At the present time, there is a good annual demand for the fruit, and any extreme variation in crop yield can be equalized to some, extent by processing the surplus in years of plenty and selling in short crop years. It is as stable as any small fruit industry could be.

The investment necessary to build a cranberry bog amounts to from $1000 to $3000 per acre, depending on the sort of bog built. Extensive planning is necessary on a project of this size. In addition, it takes persistence to overcome the unexpected reverses that come to every grower at times. Results are slow to appear, but when they do, they repay the grower well.

The manual work of the owner is secondary to his executive ability and foresight, if his planting exceeds 10 acres. He has to take advantage of local growing conditions and varied weather to get the best results. For instance, in applying sand to bogs, it is economical to put the sand on ice in the winter, but often good ice lasts only a day or two. If the grower has made plans for sanding and has protected his sand supply from freezing, he can hire trucks and power loaders for a day or two at the right time and get a great deal of work done. When the ice melts, the sand sifts through very evenly. Otherwise, the resanding must be done by shovel from a wheelbarrow.

The reason for the sanding operation is simple: A cranberry bog consisting of peat only is poorly aerated and allows roots to enter but a short distance. The runners from the original plant are somewhat like strawberry runners and should root with great freedom. The application of sand gives an aerated portion which allows the roots to enter some distance and be protected from minor droughts. Resanding covers the runners and fallen leaves, encourages additional rooting and abundant uprights all over the bog. Such a growth is very fruitful and can be harvested economically.

There are times, too, when long hours must be put on such special jobs as frost protection, the locating and controlling of insect pests, and the supervision of harvesting. These jobs are never completely finished, but extensive effort is well repaid. These are the jobs that make the business interesting and exciting.

AREAS SUITABLE FOR CULTURE

Cranberry culture is restricted climatically to a latitude north of Washington, D. C., and to those locations having an acid peat soil that can be submerged with naturally acid water at the will of the operator. This narrows the suitable localities to areas in southern New Jersey, east central New England—principally Massachusetts—northern Wisconsin, and the western parts of Washington and Oregon.

The suitable soil in these areas is extensive and a considerable industry has been established. It appears that elsewhere the desirable sites are so limited that not more than one or two bogs can be built in any one section. Anyone attempting to raise this fruit far from one of the main growing districts would have difficulty getting experienced labor and suitable tools and supplies for the work. The total acreage in cranberries in the United States amounts to about 28,000 and the annual yield is in the neighborhood of 600,000 barrels.

ATTRACTIONS AND PITFALLS FOR CITY DWELLERS

There are features of this business of cranberry growing that are particularly adaptable to the needs of city dwellers who want to return to the land, and many are attracted to it. The areas are near summer resorts and their climatic features are desirable for a vacation land. The total area is limited. The investment necessary per acre is much larger than that for the usual type of farm, and some years elapse before the first returns appear. This factor is an advantage in that it eliminates most of the competition with which ordinary farmers have to contend. Most of the work will have been done when the cranberry bog starts to bear, and a grower can then look forward to 20 years of good crops without thinking of large new expenses.

Hobby farmers usually have considerable capital or they would not be looking for a farm. Their regular income is probably secure for 5 to 10 years ahead, and that will see them through the cranberry bog's development period. A large amount of careful planning has to precede any of the field work in order to make sure that management and financial needs can be met. It is not a matter to be entered into lightly, for a new bog requires capital constantly and it is easy to spend too much for the wrong kind of development.

The new grower who tries to buy a bog already bearing fruit has a rude shock awaiting him. He has not had the experience necessary to recognize a good place from a poor one and is in no position to make a selection. Bogs with good production records are not always the best-appearing bogs, or, if they are, they may be run down. Plans for rebuilding run-down bogs have to take into consideration the partly worn-out condition of the soil, and this factor usually throws the balance in favor of starting on virgin land. Many bogs have changed owners, but few good-bearing bogs have passed from the family of the builder—and these few have commanded a price no novice would, think of paying.

Another method of entering the business is for the city man to form a partnership with an experienced grower. Usually such a plan does not actually operate as well as it would seem to do on paper, and it is extremely doubtful if it would benefit both parties in any small operation. Stock companies have been used, but often they fail because of the lack of interest in the details of management on the part of the owners. Doctors, lawyers, ministers, and school teachers generally have been unsuccessful unless they have had unusual foresight and business ability and could afford to spend considerable time learning the rudiments of cranberry culture. Surprisingly enough, when such professional men and women do go into the business and give it all it takes, they become outstanding leaders with accompanying success and responsibility.

It is rare to find an owner of a few acres satisfied, even though he could make a good living easily. After the extensive work of bog-building, it is difficult to stop further development and simply care for and harvest the crop. Many growers reinvest their earnings in more bogs, and eventually they have more than they can maintain and no reserve capital. Hobby growers, at least, might well limit their activities strictly to operations well within their resources and not get trapped in an ever-increasing business project.

CHOOSING THE SITE

The site of a cranberry bog has much to do with its future fruitfulness. Wild land is usually selected for the site of a new bog, and the land's virgin growth gives a fair indication of the type of soil. Deep peat, for instance, has a dense growth of white cedar; shallow peat is marked by a mixture of cedar, swamp maple, and pine, and Savannah soil usually has a mixture of leather-leaf, grass, and wild cranberries. Deep peat is probably the best soil type for cranberries, but it is much more expensive to clear. The Savannah type is much easier to clear and probably just as profitable during the first 10 years of use as a cranberry bog.

Section of bog soil showing peat at the bottom with a sand layer and the vines growing on the top.

The water supply should be dependable and steady, with a well-sustained dry-weather flow and as much freedom from floods as possible. A site on a small tributary of a much larger stream is often used effectively by leading water at will from the main stream across to the tributary, thus furnishing a plentiful supply of water. Of course, the stream below the bog must be cleared to furnish sufficient drainage below the planting. A site 2 to 4 feet above that of the level of a lake or pond can be used by pumping the needed water into the bog and draining back into the lake.

Except in Washington and. Oregon, water is needed for submerging the entire bog during December, January, February, March, and part of April, to protect the vines from winter-killing. The vines may be reflowed about June first for insect control. Partial flooding is also used for control of frost, which may occur during May, June, August, and September, or until the berries are harvested. During the remainder of the growing season, water should be maintained at a depth of 12 to 18 inches below the surface, depending on the amount of sand on top. In Washington and Oregon, where the soil does not freeze in winter, the bogs are not flooded, and overhead irrigation is used to some extent for frost protection.

FLOOD CONTROL

Flooding the bogs during bloom ruins the crop for the year, and flooding any time during the growing season reduces the keeping quality of the fruit. For this reason, growers usually have arrangements for by-passing surplus flood waters during spring and summer freshets. This contingency occurs possibly once in two years, but the danger is ever-present and considerable care and expense to avoid it is advisable.

The dams for water control are made of earthwork. After

clearing the vegetation from the site, a core ditch is dug through the muck to the sand underneath, and all logs and stumps encountered are carefully removed. The trench is then filled with sand, and the sand filling is continued to the desired height. A 12-foot width at the top is usually ample, and a slope of slightly more than 45 degrees is allowed. The sides of the dam are protected by turf sods, carefully placed in a single layer like so many large bricks against the sand slope. The top of the dam is usually a roadway.

The gates and trunks of the dam are made of wood or concrete and vary greatly according to the amount of water to be held and the amount to run through regularly. The requirements are that they be tight, fairly permanent, allow drainage 18 inches below the surface, and remain in place under the conditions of the location.

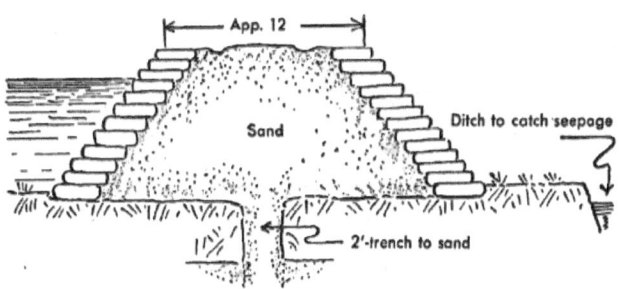

Cranberry dam cross-section, showing core ditch to sand subsoil, the turf sides, and the ditch to catch seepage.

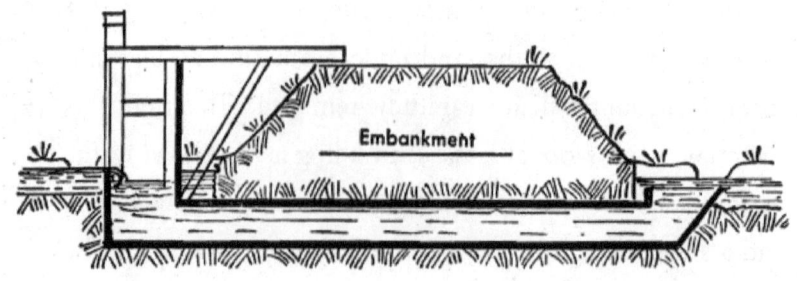

Sluice gate cross section, showing sluice entirely under water, gate away from dam, and platform at top for worker when he moves boards in and out of gate.

CLEARING THE LAND: BOG PREPARATION

The land to be used for cranberries can be flooded as soon as the dam and gates are built. Through years of experience, it has been found that the most economical method of clearing this type of soil is to burn what remains after removing any valuable timber, and then submerge the site for two growing seasons. The treatment kills all roots and prevents their growth as weeds after the cranberries are planted. However, it is quicker simply to burn off the wild growth on the site and turf the area. This is accomplished by cutting the turf in strips a foot wide and then rolling it up into convenient sizes and removing it completely.

The land can then be cleared of stumps or any other obstruction at some time during the operations, and a system of open drainage ditches installed. A marginal ditch all around the future bog and cross ditches 40 to 75 feet apart, leading from the marginal to the center ditch, usually allow for sufficient drainage.

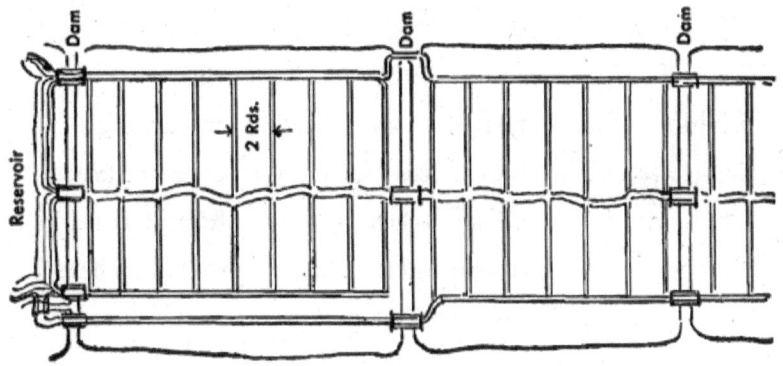

Layout of drainage and irrigation system on cranberry bogs, showing reservoir, side ditches, regular cross ditching to central ditch and dams.

The surface of the bog is leveled as completely as possible. If it is absolutely level, flooding can be accomplished more economically, but a variation of not more than a foot in a 20-acre bog is very good. It is obvious that flooding must be accomplished with a minimum of water, as there is usually a shortage of water when it is needed most. In addition, deep flooding such as would be necessary to cover an uneven bog is often injurious to the cranberry vines.

When the site has been cleared and leveled, 1 to 3 inches of coarse sand is spread evenly over the surface of the bog and the water table lowered to from 15 to 18 inches below the surface. The soil is now ready to receive the cranberry cuttings.

SELECTING VARIETIES; PLANTING; CULTIVATION

The varieties used today for planting are limited to Early Black, Howes, McFarlin, and Searls. Early Black is a good producer, resistant to false blossom disease, and an excellent market berry. Howes produces well and is an excellent keeper, but the vine is not resistant to false blossom. McFarlin is a large berry grown in Wisconsin and the Pacific Northwest, where it is a fair producer and exceptionally resistant to false blossom. It does not color well in the East. Searls is a large berry, an excellent producer, and is rapidly becoming the chief variety in Wisconsin, but it does not resist false blossom. All these varieties were selected and developed from wild cranberries at some time in the past, but now improved varieties are being developed by the U. S. Department of Agriculture and possibly some of these will soon be selected as superior.

FALSE-BLOSSOM VINE COMPARED WITH
NORMAL VINE: (1) Normal, A—unopened bud, B—open

flower, C—young fruit just after blossom has fallen, and D—young fruit. (2) Shoot showing early false blossom malformations, A—flower with calyx lobes somewhat broader than normal and the petals much shortened and broadened, B—flower with the sepals broadened and divided at the base; the petals are short and broad approaching a foliaceous condition, the stamens are somewhat shortened and abnormal and the ovary abnormal, elongated into a conical form and infertile, C—like B.

CRANBERRY VINE: Central section shows stem originally set out with a runner to each side. Leaves for any one year bunched together and the fruit occurring just under the current season's leaves. Note leaves for three years may be on a single shoot, as the one next to the right end shoot.

In selecting cuttings for planting, a grower should be sure that they are free from false blossom disease. No vines having this disease should be used; areas where vines are to be cut for planting should be selected the previous August, when the absence or presence of the disease is obvious. False blossom is recognized by its tendency to cause witches' broom, smaller leaves, and the failure of the bloom to turn down as is normal. The malformed blossom remains upright like a daisy. The foliage turns red in mid-August about a month ahead of normal foliage. Vigorously growing vines free from mixtures of other varieties, as well as from insects or diseases, should be selected for planting.

The vines are mowed like so much hay, usually with a scythe but, if large quantities are needed, sometimes with a mowing machine, preferably on the day they are to be planted. If they must be kept for a few days, they can be baled loosely and submerged in cold water. The planting distances vary from 6 to 12 inches each way, but 6 by 16 inches is a good distance, as this allows weeding with a garden wheel hoe several times during the first year. A single upright or vine in one place is better than several, for a better contact can be made with the soil and the planting can be rogued for false blossom appearing after planting.

After the first year, the vines spread over the soil, so that any additional weeding must be done by hand. Freedom from weed growth assures maximum light and root space to the cranberry, and therefore quicker growth. As the vines spread out, they root more quickly if an additional sanding of about a quarter of an inch is applied partly to cover the runners actually on the ground. Uprights 6 to 8 inches long hide the soil almost completely on a well-kept bog. The third or fourth year brings the first crop, if all goes well.

METHODS OF CONTROLLING PESTS

The most interesting operation on cranberry bogs is the fighting of economic insects by submerging the bogs and thereby drowning or washing the insects away. This is a simple treatment and, when well timed, it is usually nearly 100 per cent effective. Unfortunately, not all insects can be controlled by this method, and the time for flooding is not the same for all species. The vines fail to do well if the submergence is repeated too often. Usually, after complete control of any insect has been effected, the particular species does not reinfest the bog for several years. Thus the grower can keep most insects under control if he submerges his bogs once every other year.

Of course, some spraying must be done—it is too dangerous to submerge the bogs every time an insect pest should be treated. The judgment of the grower is important in determining whether it is safe to submerge. Fresh rain water or cool water from an active stream contains much dissolved oxygen and is relatively harmless to cranberry vines, while water stored for some time in a reservoir is apt to have insufficient dissolved oxygen and might easily cause injury to the vines. When the same water is used repeatedly in a string of bogs it may be harmless to the first bog, cause some damage to the second, and ruin the crop on the third, for the water loses its dissolved oxygen under such conditions.

Rapidly growing cranberry vines suffocate easily in warm water, especially on cloudy days. In the sunshine, the plants give off enough oxygen to keep the water fairly well supplied, so that the plants can exist over night, but under bad conditions it is best to limit reflows even with clear weather and with cool water. The length of the reflow should be limited to 24 hours if the water temperature is 85 degrees. F., but it can be extended to 50 hours if the water temperature is 65 degrees. Oxygen stays in solution with great difficulty as the water warms up.

Insects Controlled by Flooding

Four kinds of insects—the blackhead fireworm, the blossom worm, the blunt-nosed leaf hopper, and the cranberry girdler—all of which are particularly destructive, are all susceptible to control by flooding.

The blackhead fireworm overwinters as an egg laid on a leaf the previous July or August. It starts out as a leaf miner for a few days, then crawls from the mine to the terminal bud and eats it just about the time it is starting to swell. As new growth develops, they go to new tips and web leaves together, eating on the surface of the leaves. This work is completed about the first of June, but a second generation appears toward the end of June. This generation usually webs several tips together, eating all the green foliage so that the bog appears as though it has been swept by fire.

The blossom worm eats foliage to some extent before the bog comes into bloom, but its most important injury is cutting off the flowers and dropping them to the ground. Serious infestations of blossom worms will cut the entire crop of blossoms from a bog and render it fruitless in 2 or 3 days.

The blunt-nosed leaf hopper appears about the first of June on bogs where no special means is taken to control it. It does considerable damage sucking the juice from the cranberry plant. Its most serious damage, however, is in its carrying of the false blossom disease as discussed under diseases.

Cranberry girdler eats the bark from the stem of a cranberry plant at the surface of the ground and is especially severe where the vines are long and run along the surface for a distance. They do not eat below the surface of the ground. It naturally follows that considerable girdling is done. Large patches of vines up to an acre or more have been killed out by the activity of this insect where the vines are long and there is considerable protection due to dropped leaves on the surface of the bog. Bogs regularly resanded have a minimum of this type of injury, although even there the insect may girdle many plants. The millers appear during June and the heaviest feeding is done during August.

The following table gives the data necessary for the control of these insects by flooding. The dates given for flooding are based upon New Jersey conditions but apply with minor variations in most other cranberry sections:

Insects	Duration of Flood	Dates for Flood
Blackhead fireworm	48 hours	May 28 approx.
Blossom worm	12 hours	June 5
Blunt-nosed leafhopper	12 hours	June 5-10
Cranberry girdler	1 week	Sept. 25 if the crop has been harvested.

Insects Not Controlled by Flooding

Insects not controlled by flooding but held in check by late holding of winter flood are the cranberry fruit worm and the yellowhead fireworm. The cranberry fruit worms start working about the time the berries are formed and mine into the fruit, eating the flesh and particularly the seeds, leave one berry and go to another and altogether destroy 3 or more berries each during the season. They are not particularly serious where the winter flow is held as late as May 10, but on early-drawn bogs they are sometimes numerous enough to ruin the entire crop. Dusting with rotenone insecticide is the best method of control.

The yellowhead fireworm overwinters as an adult but lays eggs promptly after the bog has been drained. It feeds very much like the blackhead fireworm, except that it has in addition a third generation appearing in late August. This not only eats the foliage but also mines the fruit.

Grasshoppers are particularly obnoxious in weedy bogs and bogs not sprayed at all. They eat the berries during July and August. They are not controlled by water.

In Washington and Oregon, where flooding is not practiced, more insecticide must be used. The lack of winter flooding there allows insect growth that is not common in eastern bogs.

Spray and dust insecticides are applied from various kinds of machines, including aircraft, all attempting to put a fine spray or dust coating on the vines with as little mechanical injury to the plants as possible. It must be remembered that cranberry vines cover the ground completely and that walking or riding over them breaks off or destroys a great number. The apparatus applicable to one bog need not be the best for another of different size and shape.

Diseases That Bear Watching

Diseases affect cranberry vines, foliage, and fruit. False blossom is a virus disease which renders the plant fruitless but does not kill it: the misshapen plant remains as a source of infection. The blunt-nosed leaf hopper carries the disease. Since false blossom is incurable, the only way to check its spread is by the control of the blunt-nosed leafhopper.

Leaf drop can be caused by a fungus and corrected by spraying with a fungicide such as Bordeaux. Rot of several kinds can also be controlled with the same material. Not all bogs need to be sprayed with a fungicide to produce a good crop, but most of them would be improved by it. On a considerable number of bogs Bordeaux spraying is essential.

CARING FOR THE BOG

Frost danger on cranberries is ever present in Wisconsin and may occur as late as June and after August 20 in the eastern states. The naturally damp, low areas are very cold on clear, still nights—sometimes 20 degrees colder than the upland 6 feet higher. Reflooding, when danger occurs, is effective in raising the temperature, but its use is a matter of experience and judgment. Sometimes the filling of the ditches is all that is needed; at other times the vines must be completely submerged.

The other operations in bog management are more or less obvious. Weeding, resanding, maintenance of drainage systems, and special care to prevent forest fires are all important.

Weeds must be removed from cranberry bogs regularly—in a young bog, with a hand-pushed cultivator between the rows and ordinary handpulling in the rows. As a bog becomes covered with vines, handpulling of all weeds becomes necessary. Some growers remove all the foreign growth from bogs 2 or 3 times a year, but it is not unusual to allow some weed growth. Holding the water table so that the surface of the sand is quite dry prevents the germination of the weed seed to a great extent. The application of weed killers is not generally satisfactory.

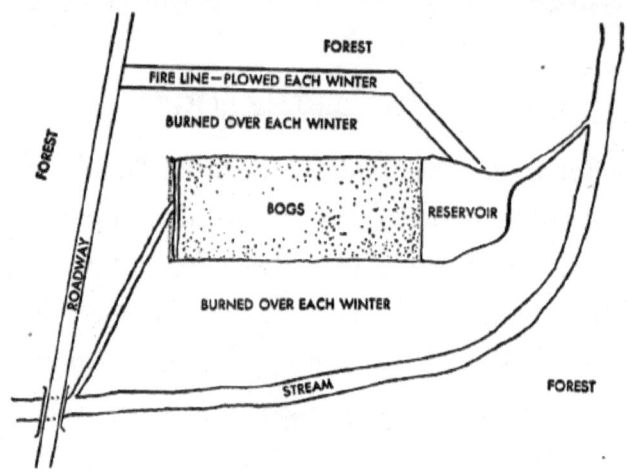

Layout of cranberry bogs along one side of a stream, showing method of taking water in above and outletting it down stream. This also shows the method of fire protection.

Bogs may be resanded every 3 to 5 years with profit. This treatment covers up the runners lying at the surface of the ground, thereby encouraging more extensive rooting. At the same time it covers up fallen leaves and any other trash at the surface of the ground and furnishes a new area of organic matter suitable for the roots to live in.

Cranberry bogs are more combustible than grass fields, and considerable care is necessary to prevent their being burned in a forest fire. Clearing out the undergrowth about the bog is a common procedure and, in addition, fire lines having no vegetation at all are often maintained at some distance back from the bog. These mechanical barriers assist greatly, but, when a fire threatens, it is necessary to have experienced men to fight the fire, as it is difficult to stop and sometimes jumps as much as a mile ahead in a stiff breeze.

HARVESTING THE CROP

Cranberries are harvested at one picking, the grower taking care to delay his harvest until all the fruit is reasonably colored. He wants to be ahead of any frost damage and yet get all the color he can on his berries. Harvest usually starts early in September and finishes before the end of October. Hand picking, scooping, and water-raking are the principal methods used, although some power machines are being tried. Many local conditions—availability of labor, type of vines, and drying conditions of the climate—enter into the selection of the method.

Cranberry and Blueberry Culture

Scooping Cranberries

As soon as possible, the dry berries are placed in the storehouse, where they are held until they are prepared for shipment. This storage period may be 3 months. Most storehouses are well ventilated and cool. In the colder sections, solid, or insulated walls are built for warmth. Usually the outside temperature is too warm, and protection against this warmth is important. Most houses can be aired well at night and closed during the heat of the day. Cold storage is used at some bogs.

The minimum storage requirement for cranberries is a roof over berry boxes piled so that air can circulate through them. Berries so stored would keep very well for a month, provided no freezing weather occurred. If berries are to be sold for canning and can be delivered within a month's period, no further preparation is needed. Many storehouses consist only of a good roof and slatted sides, leaving one-quarter of the area open so that the air can blow through freely. Berries kept after Thanksgiving need protection from freezing weather, and some storages are provided with well-insulated walls and enough heaters to keep the temperature above freezing. The most satisfactory storehouse provides for cold storage during the early part of the season and heat during the latter part.

COOPERATIVE MARKETING OF BERRIES

The marketing of the cranberry crop is much more orderly than that of most agricultural crops. Large cooperatives handle the crop almost entirely, and the grower has no personal contact with brokers or commission men. The berries are packed and graded under the brands of the cooperative and are loaded into a railroad car or truck that the cooperative has sent.

The American Cranberry Exchange is the largest shipper of fresh fruit, and Cranberry Canners, Inc., is the largest shipper of processed fruit. There are other cooperatives, and a few growers sell their berries independently.

Experienced sales management assures the grower of a fair price for his crop and almost eliminates the price fluctuations that mark the sales of some fruit. Advertising, sales studies, research in the uses of cranberries, and up-to-the-minute commercial reports are used to advantage by these organizations.

SUGGESTED READINGS

General

CRANBERRY ACREAGE AND PRODUCTION IN MASSACHUSETTS, by V. A. Sanders. (New England Crop Reporting Service, Department of Agriculture, Boston, Mass.)

CRANBERRY BOG WEEDS, by H. J. Franklin and others. (Cranberry Canners, Inc., Hanson, Mass.)

CRANBERRY FERTILIZERS, by Charles S. Beckwith. (*Circular 313*, New Jersey Agricultural Experiment Station, New Brunswick, N. J.)

CRANBERRY GROWING IN MASSACHUSETTS, by Henry J. Franklin. (*Bulletin 373*, Massachusetts Agricultural Experiment Station, Amherst, Mass.)

CRANBERRY GROWING IN NEW JERSEY, by Charles S. Beckwith. (*Circular 246*, New Jersey Agricultural Experiment Station, New Brunswick, N. J.)

THE CRANBERRY INDUSTRY IN MASSACHUSETTS, by C. D. Stevens and others. (Massachusetts Agricultural Experiment Station, Amherst, Mass.)

ESTABLISHING CRANBERRY FIELDS, by George M. Darrow and others. (*Farmers' Bulletin 1400*, U. S. Department of Agriculture, Washington, D. C.)

HARVESTING AND HANDLING CULTIVATED CRANBERRIES, by H. F. Bain and others. (*Farmers' Bulletin 1882*, U. S. Department of Agriculture, Washington, D. C.)

LATE HOLDING OF WATER ON CRANBERRY BOGS, by Charles S. Beckwith. (*Circular 402*, New Jersey Agricultural Experiment Station, New Brunswick, N. J.)

MANAGING CRANBERRY FIELDS, by George M. Darrow and others. (*Farmers' Bulletin 1401*, U. S. Department of Agriculture, Washington, D. C.)

SANDING CRANBERRY BOGS, by Charles S. Beckwith. (*Circular 371*, New Jersey Agricultural Experiment Station, New Brunswick, N. J.)

A SURVEY OF THE CRANBERRY AND BLUEBERRY INDUSTRIES IN NEW JERSEY. (*Circular 232*, New Jersey Department of Agriculture, Trenton, N. J.)

WEEDS OF CRANBERRY BOGS, by Charles S. Beckwith and Jessie G. Fiske. (*Circular 171*, New Jersey Agricultural Experiment Station, New Brunswick, N. J.)

Cranberry and Blueberry Culture

Periodicals

THE NATIONAL CRANBERRY MAGAZINE. (Published monthly by the Wareham Courier, Wareham, Mass.)

PROCEEDINGS OF THE AMERICAN CRANBERRY GROWERS' ASSOCIATION. (Published semi-annually by the Association, Pemberton, N. J.)

PROCEEDINGS OF THE CAPE COD CRANBERRY GROWERS' ASSOCIATION. (Published semi-annually by the Association, Wareham, Mass.)

PROCEEDINGS OF THE WISCONSIN STATE CRANBERRY GROWERS' ASSOCIATION. (Published semi-annually by the Association, Wisconsin Rapids, Wis.)

Insects

CAPE COD CRANBERRY INSECTS, by Henry J. Franklin. (*Bulletin 239*, Massachusetts Agricultural Experiment Station, Amherst, Mass.)

CRANBERRY BLOSSOM WORM, by Charles S. Beckwith. (*Circular 312*, New Jersey Agricultural Experiment Station, New Brunswick, N. J.)

CRANBERRY FIREWORMS, by Charles S. Beckwith. (*Circular 321*, New Jersey Agricultural Experiment Station, New Brunswick, N. J.)

CRANBERRY GIRDLER, by Charles S. Beckwith. (*Circular 314*, New Jersey Agricultural Experiment Station,

New Brunswick, N. J.)

CRANBERRY INSECT PROBLEMS AND SUGGESTIONS FOR SOLVING THEM, by H. B. Scammell. (*Farmers' Bulletin 860*, U. S. Department of Agriculture, Washington, D. C.)

Diseases

CRANBERRY DISEASES AND THEIR CONTROL, by C. L. Shear. (*Farmers' Bulletin 1081*, U. S. Department of Agriculture, Washington, D. C.)

CRANBERRY FALSE BLOSSOM AND THE BLUNT-NOSED LEAFHOPPER, by Charles S. Beckwith. (*Bulletin 491*, New Jersey Agricultural Experiment Station, New Brunswick, N. J.)

FALSE BLOSSOM, by Henry J. Franklin. (*Extension Leaflet 154*, Massachusetts State College, Amherst, Mass.)

THE SPREAD OF CRANBERRY FALSE BLOSSOM IN THE UNITED STATES, by Neil E. Stevens. (*Circular 147*, U. S. Department of Agriculture, Washington, D. C.)

STUDIES ON CRANBERRY FALSE BLOSSOM DISEASE AND ITS INSECT VECTOR, by Irene D. Dobroscky. (*Contributions from Boyce Thompson Institute, Vol. 3, No. 1*, Boyce Thompson Institute for Plant Research, Inc., Yonkers, N. Y.)

BLUEBERRY CULTURE

CHARLES A. DOEHLERT

Charles A. Doehlert, in this chapter, draws on 13 years of intimate association and work with the men and women who established and developed the still-young cultivated-blueberry industry. What has since been done in other sections of the United States in the growing of this specialty crop has been patterned in no small measure on the practices evolved in the Garden State. As associate in blueberry (and cranberry) research for the New Jersey Agricultural Experiment Station, Mr. Doehlert has contributed significantly to the industry's sound development, particularly in the use of fertilizers, methods of tillage, and pruning. With Charles S. Beckwith, he shares the credit for developing a method for controlling the blueberry fruit fly, an insect that seriously threatened the industry. He is a graduate of Rutgers, where he majored in horticulture.

IN GENERAL, blueberry growing for profit is a business in itself. The blueberry plant requires an unusual type of soil, so the man who plans to farm ordinary loam soil will do best to cross blueberries off his list of small fruits. This plant is not a neighborly companion to be grown in a few rows alongside the raspberries, blackberries, or strawberries.* However, if the farmer has land where other acid-tolerant plants such as white cedars, cranberries, red maples, wild huckleberries, or wild blueberries are growing wild, he may wisely take time to consider the rewards and trials of blueberry growing. The fruit is popular and the market so far has been good. The problems, risks, and investment at stake are considerable.

Anyone embarking on farming as a new experience will quite possibly be rewarded with both profits and pleasure, if he concentrates on blueberries in a favorable location instead of trying to combine that enterprise with some other crop, to the disadvantage of both.

In many cases, established cranberry growers have taken on blueberries as a side line. In general, this has turned out to be simply the addition of another business. Somewhat more labor can be kept the year round for use in the winter for blueberry pruning and in the fall for cranberry harvesting. But many growers have found that competition for labor at other times of the year is very apt to work out to the serious detriment of one of the two projects. Well-rounded experience with one of the two crops would certainly be required before any sound plans could be laid for undertaking the other.

Average Run of Cultivated Blueberries (4/7 natural size)

Average Run of Wild Blueberries (4/7 natural size)

The person who has an established income, some spare capital and time on his hands, and a piece of blueberry land located conveniently near at hand is in a favored position for starting out in a small way and learning by the trial and error method. This is the way most successful growers have gotten into the business and remained there.

ATTRACTIVE FEATURES OF BLUEBERRY CULTURE

Blueberry culture may be conducted as a small business of 10 to 20 acres, in which the owner can be active the year round with a considerable variety of outdoor and administrative work. If he has competent help, he may manage well with short hours in the field, except during 5 or 6 weeks of harvesting and during periods of frost and pest control. Or he may arrange to leave his field in charge of a subordinate during the slack season of September and October, and again during January and February, if pruning is well under control.

At the other end of the scale, the owner or manager may depend largely on employees for the manual work and find a considerable outlet for his administrative skill in directing operations. But in any case, the owner's frequent or daily presence for 8 months of the year and a complete familiarity with the manual work are usually essential for success.

There is a widespread public interest in cultivated blueberries as a food. This is due to their delicate, agreeable flavor, their relative scarcity, and the long-standing and general popularity of "huckleberry" pie. The superior flavor and appearance of the cultivated berries, plus the cleanliness and uniformity of pack, makes them popular with the cook as well as the epicure. This popularity is being shared by the canners and "quick-freezers" who take a considerable portion of the crop. The increasing difficulty in getting wild blueberries that pass the standards of the Federal Pure Food laws has been a factor in maintaining a strong market for cultivated blueberries.

There are many opportunities for the application of imagination, inventiveness, and good management in increasing efficiency, reducing costs, and improving the marketed product.

Competition is fairly limited due to the plant's soil requirements and unusual methods of culture. It will probably continue to be limited by the newer pests not yet under control.

Up to the present, blueberry growing has been one of the more profitable lines of farming.

WHERE BLUEBERRIES ARE GROWN

Since blueberries thrive best in an acid, sandy soil well supplied with muck or peat, the prosperous blueberry sections are not those where general farming has become well established.

The novice setting out to observe blueberry culture first-hand is usually surprised to find himself in a generally undeveloped section, where the soil is considered poor and the chief ornaments to the landscape are pines, cedars, red maples, holly, and scrub oaks. The streams are clear and often pure, but the water has taken on a brown color from the peat of cedar swamps. Human habitations are few and far between.

If this type of country is pleasing to the future grower's individual tastes, he probably finds himself considering how this location and the facilities for transportation meet his requirements for a home, public schools, business contacts, and social life. Frequently, the blueberry grower chooses to live in the nearest town which provides suitable community advantages and access to his not-yet-abandoned interests in the city as well as to his plantation in the woods.

Adapting oneself to the conditions of an established blueberry section is of advantage in that efficient marketing facilities have been developed and local labor is experienced in the business. There are at present three regions where the industry has developed extensively: southern New Jersey, southeastern North Carolina, and southern Michigan. Southern New England and parts of New York State offer possibilities as areas for future expansion.

INVESTMENT AND RETURNS

It takes 3 to 5 years to bring a blueberry field into profitable production after the land has been cleared for plowing. This period of small returns but accumulating expense is a large factor in the cost of a field. The wild land itself should be bought cheaply, although it may be necessary to purchase considerable adjoining acreage to obtain the piece which is suitable for planting. Land bought at $25 an acre has involved an investment of as much as $1,000 per acre in cash and labor by the time the plants were full grown and bearing heavily. About 10 acres are generally required to make a business large enough to keep unit costs of production at a low figure. A plantation of such size is large enough to keep the owner or an employee present and busy on the place the year round, and this watchfulness is of considerable advantage as a protection from such hazards as marauders, fire, and flood as well as in the prompt detection of pest attacks.

In a season when the average selling price was 29 cents a quart, all costs incident to picking, packing, shipping, and selling were computed at 15 cents. Costs of production, including taxes, interest on investment, employer's liability insurance, and supervisor's time at the field at laborer's wages, were computed at 8 to 12 cents a quart. On a 50-acre plantation where the production cost was estimated at 8 cents a quart, the yield was about 75 bushels per acre. On a 10-acre plantation where the production cost was estimated at 12 cents a quart, the yield was about 100 bushels per acre. Both of these plantations therefore yielded a substantial profit over and beyond overhead and operating costs plus wages for the owner's labor in the field. The difference in production costs between the two plantations was due, not only to size of operations, but also to the fact that fewer pests and stricter management economies favored the plantation with the lower unit cost. Both were located on good soil.

FUTURE OF THE INDUSTRY

Blueberry culture is still a new industry. After an experimental period of 30 years, the first commercial shipments of cultivated berries were made in 1919. Blueberries have a wide appeal and are still attractive as a novelty. Compared with the wild fruit, the cultivated is much more attractive in appearance. It is also easier to prepare for use by virtue of its freedom from trash, dirt, and under-ripe or spoiled berries.

Although many other northeastern cities receive cultivated blueberries, the bulk of the crop is still sold in New York and Boston. The market for this fruit was surprisingly stable throughout the decade of the 1930's. The Blueberry Cooperative Association, organized in New Jersey in 1927, has been the means of distributing the berries in accordance with demand and without glutting any one market. Hence there have been no serious price slumps with large losses to the growers, such as occur with most perishable fruits. It was fortunate for the infant industry that it was developed by growers already experienced in cooperative marketing. As the North Carolina and Michigan fields came into bearing, those growers, too, joined the Blueberry Cooperative Association.

In recent years, as the output has rapidly increased, the Cooperative has made arrangements for canning and freezing during the peak of the harvest season. This greatly extends both the season and the area for the consumption of the fruit and helps prevent forced sales at prices below cost of production when the berries ripen faster than established markets can readily absorb them.

As the quality of cultivated blueberries has become known, consumers' standards have been raised. Each year competition with wild berries becomes of less importance. Altogether, the popularity of the fruit and the large, untapped markets still help to make blueberry growing one of the more attractive farming ventures.

PROBLEMS THE GROWER FACES

There is, however, a reverse side to the picture. There are ordinary hazards, such as a severe winter freeze or a spring frost, or extremes of rainfall or drought during the fruiting period, which may ruin a year's crop. More serious than these are the inroads of new diseases or pests which may be very destructive before control measures are discovered. So far this type of loss has been disturbing and costly for a period, but ways and means for control have always been found by State and Federal research workers in time to avoid disastrous losses.

Perhaps the most unpredictable hazard for the future lies in marketing. As any industry expands geographically and in numbers of growers, cooperative marketing becomes more complicated in management. If this type of marketing ever ceases to be the dominant factor in the blueberry industry, it will be anyone's guess as to whether prices will pay for the cost of production. In such an event, the least efficient growers will be forced out of business, and the most resourceful will find ways of selling a better product at premium prices or of producing at a lower cost per unit.

LOCATING THE BLUEBERRY FIELD

Since established blueberry fields are rarely offered for sale the would-be grower must learn how to recognize good, undeveloped land; otherwise he may expend years of labor and much capital on soil that cannot possibly produce profitable crops.

Good blueberry soil is usually a black, acid, sandy soil with a considerable content of peat or muck. The best results have been obtained on soils ranging in acidity from pH 4.0 to pH 4.8. The sand is valuable in keeping the soil loose and open, thus aiding drainage and aeration. The peat or muck is responsible for the very acid condition of the soil. It furnishes elements of fertility and acts as a reservoir for moisture. Coarse sand and rough, partly-decomposed muck, commonly called peat, are much more favorable than fine sand and fine, well-decomposed muck. The coarseness of these materials aids in the removal of excess water, thus allowing for the entrance of air into the soil while it is still moist and favorable to root growth. Below the top layer of muck in virgin soil, is found a layer of white sand, and beneath this a hardpan. Other soils will produce blueberries, but this is the typical and most dependable type known at present.

A layer of peat 4 inches thick (exclusive of any top duff of loose, brown, unrotted debris interlaced with the roots of wild plants) will make a good blueberry soil. Such deposits of peat may be recognized in wild land by the trees and shrubs growing on it. White cedar, red maple, wild blueberries, cranberries, and leatherleaf are reliable indicators.

Good Drainage, Essential

The next important requirement is that it be possible to drain the land. The grower should be able to maintain the soil free of standing water—generally to a level of 18 inches below the surface. In wet seasons, therefore, there must be some effective outlet for removing standing water. The importance of this point can scarcely be overemphasized. There are few troubles more serious or common in blueberry growing than injury from a waterlogged soil, regardless of whether the season be summer or winter.

Protection from drought is also necessary. Abundant peat in the soil is the first line of defense, since peat holds large amounts of water a surprisingly long time.

If there is enough ground water to maintain a flow through the main drainage ditch in the driest weather, or if a near-by stream can be tapped to fill the ditches in time of drought or danger of frost, a means is presented of getting a large crop in seasons when the average field is below par, and premium prices are apt to be paid.

Temperature Limitations

Information about the customary minimum temperatures in winter and spring is a help in appraising a contemplated site. Winter temperatures of minus 20° F. are destructive to blueberry buds and indicate territory unfavorable to blueberry culture. The likelihood of these temperatures may be ascertained for a particular section from the State office of the U. S. Weather Bureau. Bodies of water in the neighborhood often make local areas safe in a region which is, on the whole, subject to damaging freezes. In a district where wild blueberries generally thrive, there may be frosts in April, May, and June that kill the buds, bloom, or fruit. These frosts will occur chiefly in spots with especially restricted air movement.

The fruiting record of the wild bushes on the land being considered is a practical indicator of frost conditions. Native pickers or near-by growers possess this information. Cold air cannot drain away from a site surrounded by higher land; such a site must be avoided. If killing spring frosts are frequent and the site is enclosed by woods, study the possibility of getting a protective air movement by clearing a strip around the field. There should be a practical opportunity to make this strip 100 feet wide. If a pond or lake can be created at the low side of the field (on the high side it would cause harmful saturation of the soil), an existing frosty condition might be eliminated. A large planting, such as 20 or 30 acres, will be less likely to suffer from spring frosts than a small enclosed patch of 5 or 10 acres.

Labor Supply

The available supply of local labor should also be considered before purchasing land. If the owner will not be devoting his full time to his crop, he will probably want a working foreman who will live on the place. This foreman should be somewhat experienced or else capable of being trained. His duties will be that of watchman and leader in the work of pruning, spraying, cultivating, and fertilizing, and in the supervision of picking and packing. Picking and packing are done chiefly by women and youths, who are more nimble-fingered than most men.

CLEARING AND PREPARING THE LAND

The largest original item of expense is clearing and preparing the land, if wild land is the starting point. An abandoned cranberry bog which permits of satisfactory drainage offers an ideal escape from the job of land clearing. In using such land, a grower can afford to spend more on the original purchase price. Occasionally, a piece of plowed, abandoned land may be secured. Land that has been growing ordinary farm crops for any length of time is rarely desirable; among other things it is not apt to have sufficient acidity. In using wild land, that covered by leather-leaf (*Chamaedaphne calyculata L.*) may be most cheaply cleared and plowed, without requiring cutting of trees and pulling of stumps. Plowing should be 6 inches deep or deeper, depending on the depth of peat present. Some sand should be turned up. A caterpillar tractor and a large plow are usually needed, because of the great numbers of tough roots to be cut and the softness of the terrain.

The field is then kept fallow by cultivation for at least a year before planting, in order to rid the soil of root-feeding grubs.

The period for these two phases of the work allows two years' time for the growing of plants for transplanting. For example, clearing the land for the main plantation and setting cuttings in the propagating bed may start at one time. During the same spring, a small piece of ground should be plowed for the "nursery," which should be kept fallow for the balance of the year while the cuttings are rooting in the propagating bed. The land for the main field will probably not be ready for plowing until late in the same season. This land will be fallowed through the second season while the small plants are making growth in the nursery.

PROPAGATION OF BLUEBERRIES

Most blueberries are propagated from hardwood cuttings. These are set out when the buds on the field plants swell, which is soon after April 1 in New Jersey and other northern sections, and in late February in North Carolina. The wood used is that portion, on sturdy fruiting twigs, just below the fruit buds. The cutting wood may be taken during midwinter or later and stored in cool, damp, well-ventilated moss. The gathering of cuttings may be delayed as late as the time of setting, which precludes loss from mold in storage but concentrates the work into a very busy season. The most common and successful type of propagating bed is set up on especially well-drained ground and shaded by a lath shelter which cuts off about 60 per cent of the direct sunlight. A mixture of sand and peat is used for the growing medium. Watering must often be done daily, with particular care to avoid any excess. Small amounts of liquid fertilizer may be used to advantage. Probably more cuttings are killed by over-watering than by drought.

Blueberry cuttings are rooted in partial shade. Laths spaced about 1 1/2 inches apart let in ever-shifting bands of sunlight and in addition protect the delicate cuttings from drying winds. This house is built with woven fencing, but plain lath can be used just as well.

CARE OF NURSERY

Continuous vigorous growth by the young plants is highly desirable. The most efficient way of obtaining this in the year following propagation is to move the plants from the propagating bed to nursery rows. This is done in the spring as soon as the soil can be worked.

Use the best ground for this purpose and do not re-use a previous blueberry bed. A portion of the main field prepared as previously described is suitable if the location is convenient for frequent attention.

The plants are set 8 to 10 inches apart in the row and the rows are usually 18 inches apart to allow for cultivation by wheel hoe. All fruiting tips should be clipped off to promote the growth of a strong bush.

Two light applications of fertilizer are advisable, the first one as soon as the plants start their second flush of foliage.

VARIETIES OF BLUEBERRIES

Each of the cultivated varieties has been carefully selected from among thousands of seedlings. Qualities of size, flavor, and attractive blue color (as well as productiveness and growth habit of the bush) were carefully weighed and considered before these varieties were named. Each variety has been chosen as pleasing to the taste and represents what would be a rare find in the wild forms. Many growers consider the flavor of Cabot too flat, but Cabot is showing promise of being a well-flavored berry for freezing. Pioneer is recognized as the hardest variety to grow.

It is impossible to state which of the 15 important varieties will be most desirable during the next 10 years. Changes in market requirements and pest hazards are to be expected in an industry as new as this one. Such changes suddenly alter variety values. Fortunately, 20 years of experience has already eliminated a half-dozen varieties which will not even be mentioned here. Subject to local conditions, all the varieties named in this discussion are somewhere worthy of a place in a blueberry field.

Early and late varieties are valued for the price premiums they bring, while midseason varieties must be very productive to become popular with growers. Open clusters, ease of separation from the stem, and a tendency to ripen all fruit in a short season are important qualities that reduce the effort of harvesting. Berries of a light, bright-blue color and large size are attractive to the buyer. A tough skin and a small picking scar (the wound made where the berry breaks off from stem) promote good keeping quality in transit to market. Ease of pruning, resistance to disease, and general thriftiness of growth are bush qualities of importance to the grower.

Canners, freezers, and ice cream manufacturers do not yet have the necessary information for requesting certain varieties by name. In the next few years they will come to the point of designating the definite varieties they want.

In each column below, the 15 important varieties are arranged in order of their fruiting from early to late in the season:

Commercially Important	Of Proved Value but Limited in Acreage	Important New Varieties Not Yet Extensively Cropped
Cabot	June	Weymouth
Rancocas	Scammell	Dixi
Pioneer	Stanley	Atlantic
Rubel	Wareham	Pemberton
Concord		Burlington
Jersey		

The six varieties listed in the first column produce the great bulk of the crop. Those in the second column have been grown rather extensively on a variety of soils in various regions and are known to be good, although they have not become so prominent. The varieties in the third column are fairly new. Since they have been selected most recently over and above the other varieties, it is very likely that they will prove to be preferred varieties.

It is advisable to grow several varieties for the sake of cross pollination and a succession of picking. The actual combination chosen depends largely upon the region in which the plantation is located. The State Experiment Station workers and the experienced growers can give valuable assistance in choosing varieties for particular localities.

DRAINAGE AND IRRIGATION

As previously stated, drainage is a prime essential for a healthy growth of blueberry bushes. Blueberry roots require plenty of moisture, but the soil must be open and well aerated at the same time. After a storm, surface water should disappear within a day or two. Generally the normal water table should be about 18 inches below the surface of the soil. Low spots in the field that do not drain off promptly must be underdrained. Because blueberries are often found in swamps, it is a common notion that wet, soggy ground is desirable. A brief, painful experience where such land is used will dispel that idea!

The winter and early spring following plowing is a good time to check up on drainage faults. Then necessary ditches may be dug and tiles laid so that the drainage system will be well understood and functioning before the plants are set in the field.

During dry spells, crops can be greatly improved if the drainage ditches and tiles are blocked and flooded for irrigation purposes. Only a few growers are using overhead irrigation so far. It is, naturally, a great advantage if wisely used.

FROST CONTROL AND WINTER FREEZES

The best blueberry soil is normally in a natural depression. On a still, clear night the cold air settles in such a depression. Very often the field is surrounded by woods which stop wind movements and aid the concentration of cold air upon the field. Severe spring frosts will occur in blueberry fields when conditions are not at all serious for farmers in general. These frosts may kill fruit buds, developing bloom, or newly-set fruit.

The simplest, and usually the only, protection available is that afforded by clearing off the surrounding woodland to aid air circulation.

When the plants are dormant, the temperature must drop to about 20 degrees below zero before fruit buds are killed, but a couple of hours of such exposure on a single night will be destructive.

EARLY CARE OF THE FIELD

The bushes are usually planted 4 feet apart in rows 8 feet apart. This calls for 1,360 plants. On particularly fertile ground, the distance in the row should be increased to 6 feet. Spring planting is customary, although not absolutely necessary. Plants that have been in the nursery only one year are preferred. A plant that remains in the nursery for two years has had its growth checked.

The fruiting tips should again be clipped off before the fruit buds start to break. This is to continue development toward a large, vigorous bush. A light application of fertilizer (200 pounds per acre) is needed when the plants have become well rooted, probably 6 weeks after planting.

The ground should be kept free of weeds. Frequent shallow cultivation around the young plants stimulates good bush growth.

During the first winter in the field, the pruning consists only in removing the weakest wood (i.e., short, slender twigs)—an operation which, incidentally, prevents overbearing and allows the development of a strong top. Strong bushes may be pruned to bear a small crop; the rest should be pruned for another season's development without fruiting.

FERTILIZER FOR BLUEBERRY CROPS

With the use of fertilizer, blueberry yields can easily be doubled. Fertilizer is applied when the blossoms first open and again a month or 6 weeks later. A 7–7–7 formula is recommended, at the rate of 300 pounds per acre for each application for full-grown plants. It should be broadcast over the soil occupied by the roots, an area which usually corresponds to the spread of the tops for the first 5 or 6 years. After that, the root-spread will be greater if tillage is not too deep. The fertilizer should be worked into the soil by machine tillage and raking. Heavier applications can be used, but these should be based upon the operator's experience with the particular field involved. On light, sandy soil, the plants may be easily burned with too much fertilizer, and on very heavy soil care should be taken to avoid excessive vegetative growth.

A good substitute for the 7–7–7 formula when this is not

available (a situation prevalent "during the second World War) will be a 4–10–5 mixture applied 600 pounds per acre (for full-grown plants) at the time of opening of bloom. In addition, 300 pounds of oilseed meal or tankage (approximately 6 per cent nitrogen in either one) should be worked into the soil about 4 weeks after applying the 4—10—5.

Fertilizer should not be applied during a drought. Addition of soluble chemicals to a dry soil has the same effect as further drying of the soil.

The bush on the right is well fertilized and bears about 3 quarts of berries a year. The bush on the left, exactly 4 feet away and planted at the same time but never fertilized (except for what it can steal from its more fortunate neighbor), bears a mere pint or a pint and a half. The fertilized bush is 5 feet tall with a spread almost as great.

PEST CONTROL VITAL TO SUCCESS

Until 1936, blueberry growers did not need to bother with regular spraying or dusting. Beginning with the necessity for fruit-fly dust that year, this good fortune came to an end. Other dusting and spraying measures are used as occasion demands, but it is certain that there will be an increase of such operations. Some growers have had to adopt annual measures against weevils. Deer have been a pest since the beginning of the industry. New pests, for which the cures are not yet known, will undoubtedly keep the struggle for profitable production from becoming mere routine.

The airplane has been the most useful means of applying insecticides. Its use, of course, is suited to large fields. Small growers with adjoining plantations can use the same plane to advantage by cleaning up their hedgerows and pooling their orders for plane service. Under peacetime conditions, New Jersey growers have secured airplane dusting service at $2.00 an acre, exclusive of materials.

A small power sprayer, developing 300 pounds pressure, with a 50-gallon tank, is a valuable adjunct to blueberry growing. It should be built narrow, 3 feet wide at a level 2 feet above ground. Low rubber-tired wheels may extend out slightly below this height.

The blueberry fruit fly has been readily controlled by two dustings a year. Although the autogiro is well adapted for such dusting, most of the commercial work is being done with the conventional plane.

The Blueberry Fruit Fly is one of the most serious pests of this crop. The fly is about one-fifth the length of the common housefly. Its grubs can render much fruit unsalable.

Successful pest control demands close attention to many details. In order to present a general view of this part of the business, the methods applying to the most important pests will be briefly summarized:

Blueberry Insect Control

Insect	Control Methods
Blueberry Fruit Fly	Derris dust, applied preferably by airplane, twice each season. *This is needed in practically all fields north of Washington, D. C.*
Cranberry Weevil	Clean cultivation. Keeping field margins clean and adjoining brushland properly burned off is a good preventive. Spray or dust before buds open. Not required in all fields. When present, weevils are usually fought in a portion of a field rather than throughout the planting.
Cranberry Fruitworm	Remove and destroy prematurely ripening fruit. If plantings of early fruit are large enough to justify the use of an airplane, two dustings with pyrethrum may be used with profit.
Blueberry Stem Borer	Break off wilted tips of the stems attacked in June and July.

Blueberry Stem Gall   Collect galls during pruning and destroy by burning.

Putnam Scale   Removal of old, weakened canes in pruning keeps scale at a minimum. For occasional severe infestations, use a dormant spray of lime-sulfur or miscible spray oil.

Blueberry Disease Control

Mummy Berry   Occasionally severe, in which case the ground under the bushes should be swept or raked to dislodge "mummies" soon after they start to germinate. Mummies are the dried-out, shriveled berries that fall to the ground and in which the disease fungus "carries-over" during the winter.

Deer Pests

In early spring, forage for deer is still scarce and the swelling fruit buds of blueberry plants are eagerly devoured. Where deer are plentiful, 6-foot woven wire fencing has usually been the only successful protection. Electric fences are on trial and in some cases have proved effective. Sound, odor, or light devices for frightening the deer away have not been successful.

New Pests Not Yet Under Control

The mite is an almost invisible, insect-like creature which

hides and feeds inside the buds. It is most destructive in North Carolina and in 1937 was conspicuously active. Since then, however, natural forces have reduced its prevalence, so that at present it is not greatly feared by growers. Attempts to control it by spraying have failed so far.

Stunt is a dwarfing disease which has recently been found scattered through the New Jersey and North Carolina fields. The total number of plants so far affected is small, but in a few fields the losses are serious. When stunt once becomes evident in a plant, that plant may be expected to be permanently out of production in 3 or 4 years. The disease is caused by a virus. At present, therefore, it is extremely important to propagate blueberries from only healthy plants.

Canker is a fungous disease which is very destructive in the North Carolina fields which have been established for 10 years or longer. The branches of infected plants become covered with a warty growth and in a few years the plant becomes unproductive. No control measure is known. Probably resistant varieties will be found.

TILLAGE AND MULCH

Tillage is valuable for both the control of weeds and the aeration of the soil. The blueberry is shallow-rooted, and tillage 3 or 4 inches deep can seriously hinder the spread of the root system. But cultivating 1 or 1 1/2 inches deep close to the plants is beneficial. A double-disc cultivator for blueberries has been designed by the author in collaboration with S. L. Allen Co. of Philadelphia. It is particularly useful after the plants have made a couple of years' growth in the field and ordinary tools can no longer be used to cultivate close to the plants. A low, spring-tooth harrow mounted on runners is also very effective and can be used on new land badly littered with roots and trash. The plants grow well on mounds, but these require more labor for maintenance and weeding than relatively flat culture. Where satisfactory drainage can not be otherwise secured, maintaining a mound on the rows is a useful last resort.

Shallow discing is the most popular form of tillage. Even when the bushes are large and nearly meet across the row, these discs can reach well underneath.

Clean cultivation improves bush growth and the yield of fruit. It also reduces insect infestation. So far, no suitable cover crop has been found for maintaining or increasing the organic matter in the soil. Mulch has been beneficial on some upland plantings. It is probably most useful on loamy soil, because it permits surface rooting without danger of drying and with the benefit of abundant aeration. Mulch has been very helpful to gardeners who fruit a few plants for home use. Building up an artificial blueberry soil is usually a requisite for success in upland gardens.

PRUNING AND TIPPING

Pruning is the largest single item of cost in maintaining vigorous, highly-productive plants. It is also one of the three most effective operations for securing large yields, the other two being water control and the use of fertilizer.

Pruning is begun as soon as the leaves drop in the autumn and on many plantations is an all-winter job. An early, intensive start is advisable, since the work slows down badly during cold and windy weather and nothing can be done in wet or snowy weather.

On certain days unsuited for pruning, the brush may be gathered from the ground and removed from the field. This brush is burned under proper authorization from the local forest fire warden. A moderate-priced shredder or cutting machine for reducing the prunings to 1- or 2-inch bits would be a boon to blueberry culture. The machine should be narrow enough to pass down the rows of bushes, so that the prunings can be loaded into it as it moves along. This would eliminate the irksome job of carrying brush out of the row and would, at the same time, return fertility and organic-matter to the soil.

Blueberry pruning is simplified by the prominence of the fruit buds, which occur entirely on the ends of the latest wood-growth made. The fundamental principles of pruning are:

(1) The largest berries are borne on the stoutest shoots.

(2) Small frail twigs bear poorly (with a few exceptions, notably on Concord).

(3) Blueberries tend to overbear and the number of fruiting twigs must be reduced to avoid weakening the bush and to allow sufficient development of strong new wood for the succeeding crop.

(4) Some old main canes must be removed periodically to let enough light into the bush for adequate development of renewal canes, which are the only means of rejuvenating the bush.

Soils, water supply, varieties, inroads of pests, and general care vary so much that an exact general standard of pruning cannot be prescribed. As the grower learns the effect of the four principal factors in his particular field, he will know whether he must prune for a moderate crop or whether he may prune for a heavy one. Good soil, adequate moisture, and enough fertilizer will make heavy pruning unnecessary and will permit the harvesting of large crops. During the first 2 years of development, when there is no pruning to do on his own field, a new grower should make it a point to do considerable pruning for others, so that he may develop the necessary judgment and proficiency.

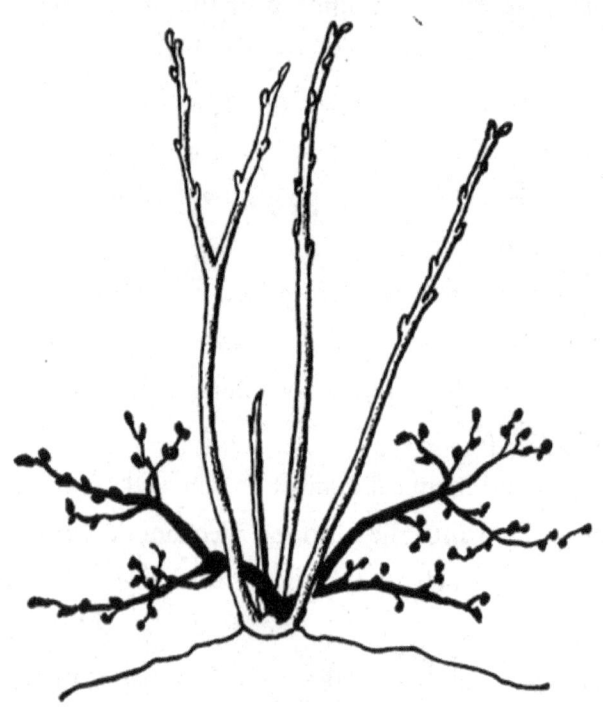

Blueberry plant after one year in the field. Branches shown full black to be removed in pruning. (Drawings by S. Coville)

Same blueberry plant after two years in the field. Branches shown full black to be removed in pruning.

Heavy pruning produces the finest, largest berries, but it also reduces the total crop. With extremely heavy pruning, the premium commanded by the higher grade fruit cannot make up for the loss of volume. Very light pruning produces a large crop, but at a serious disadvantage. With very light pruning, the berries are apt to be small, or even worthless if the weather is dry at ripening time. At the same time, the wood produced for the next year's crop will be weakened and the bush will be more susceptible to scale and cane blight. It may take two seasons of skillful pruning to restore such a bush to its original productiveness.

Tipping

On Cabot and Pioneer, tipping or thinning of fruit buds is usually advisable in order to prevent overbearing. In most seasons, the larger twigs of Cabot will produce a string of 7 to 10 or more fruit buds. Pioneer is somewhat less prolific. Better fruit and much better foliage-development will be obtained if these strings are clipped back to a remainder of 4 or 5 fruit buds for Cabot and 3 for Pioneer. This should be done after pruning is complete and during the last 2 or 3 weeks before the buds swell and break open. If protection against deer and frost is unsatisfactory, the grower may choose to gamble with too many fruit buds rather than a final supply of none at all.

HARVESTING AND PACKING

This part of the season's round is a busy time. For 6 weeks there is usually a full day's work every fair day. Harvesting begins in late May in North Carolina, about June 20 in New Jersey, and in mid-July in Michigan.

Any one bush is usually picked at 5- to 10-day intervals. Some varieties can be completely harvested in 3 pickings, others require 5 or 6. At the peak of the season, some long days of picking with night hours to finish the packing may be necessary.

Pickers are paid by the pint or quart and must be supervised closely to insure that the right bushes are picked, that care is taken to get all the ripe fruit, without stems, trash, or unripe berries, and that the picked berries are not unduly exposed to the sun.

Packers must maintain a uniform, tight pack, be watchful for the correct grading of the fruit, and reject dirty, bruised, or trashy fruit.

Above—Two fruit buds and two leaf buds as they appear in early spring. Below—same twig as it would appear at fruiting time.

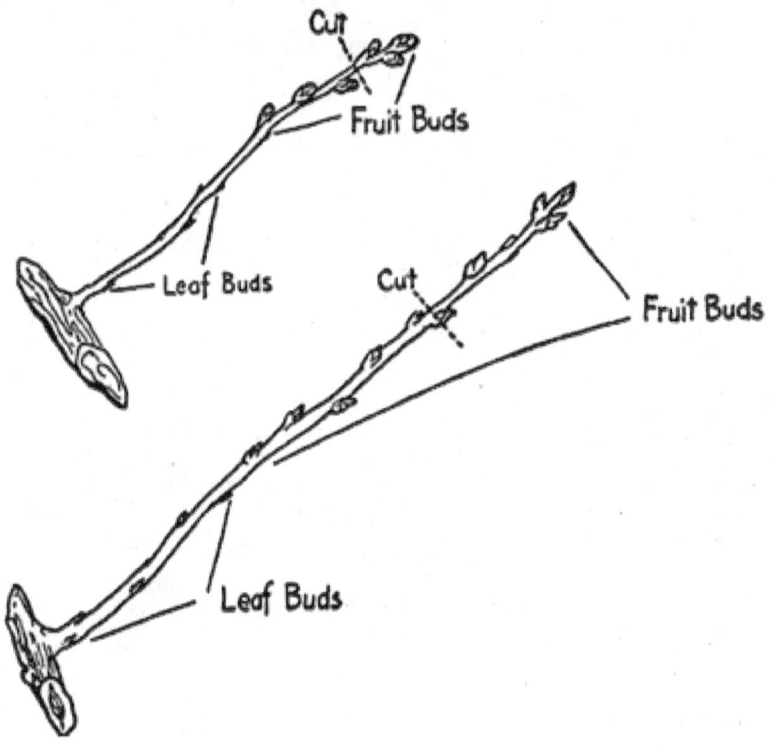

Lateral sprout of Pioneer and Cabot, showing amount of tipping-back needed. (Drawings by S. Coville)

The manager must also keep the harvest operations moving briskly enough to avoid the necessity of picking overripe berries. Cooling the berries after picking greatly improves their shipping quality.

MARKETING BLUEBERRIES

Theoretically, a grower might market his own berries either to roadside markets, other large outlets, or through a commission merchant. Practically, his time at his field is needed so urgently that he is better off if he has a trustworthy, expert marketing agency. The Blueberry Cooperative Association has met this need and has, more than any other one factor, kept the blueberry industry on a stable keel to the advantage of members and non-members.

As fast as the pickers bring their trays of pint or quart cups to the packing shed, they are sorted for grade. There is enough fruit

exposed to view to make it unnecessary to "pour" or disturb the berries for the purpose of grading. The packer places the appropriate marker, or "seal," on the face of the pack, deftly covers it with a cellophane wrapper, and places the box in a shipping crate. The girl with the long needle poised in mid-air has spotted a defective berry. She is about to "spear" it—much easier than trying to pick it out with her fingers.

By its orderly marketing and avoidance of market gluts, it has been possible to sell blueberries at a fair price all of the time. As the total crop has increased, the new outlets of canning and freezing have been developed. These extend the area of the nation's consumers that can be served and makes the product available and known the year around. This Association has made economical group-purchasing easy and practical. It has also quickened the adoption of new methods for the control of pests to the advantage of every grower.

SUGGESTED READINGS

THE ATLANTIC, PEMBERTON, AND BURLINGTON BLUEBERRIES, by George M. Darrow. (*Circular 589*, U. S. Department of Agriculture, Washington, D. C.)

BLUEBERRIES IN THE GARDEN, by C. S. Beckwith. (*New Jersey Agriculture March 1940*, New Jersey Agricultural Experiment Station, New Brunswick, N. J.)

BLUEBERRIES UNDER MULCH, by J. Harold Clark. (*New Jersey Agriculture July 1936*, New Jersey Agricultural Experiment Station, New Brunswick, N. J.)

BLUEBERRY CULTURE, by Charles S. Beckwith, Stanley Coville, and Charles A. Doehlert. (*Circular 229*, New Jersey Agricultural Experiment Station, New Brunswick, N. J.)

BLUEBERRY CULTURE IN MASSACHUSETTS, by John S. Bailey and Henry J. Franklin. (*Bulletin 317*, Massachusetts State College, Amherst, Mass.)

THE BLUEBERRY IN NEW YORK, by G. L. Slate and R. C. Collison. (*Circular 189*, New York State Agricultural Experiment Station, Ithaca, N. Y.)

BLUEBERRY TILLAGE PROBLEMS AND A NEW HARROW, by Charles A. Doehlert. (*Bulletin 625*, New Jersey Agricultural Experiment Station, New Brunswick, N. J.)

THE CULTIVATED BLUEBERRY INDUSTRY IN NEW JERSEY, 1939, by Harry B. Weiss. (*Circular 311*, New Jersey Department of Agriculture, Trenton, N. J.)

THE CULTIVATION OF THE HIGHBUSH BLUEBERRY, by Stanley Johnston. (*Special Bulletin 252*, Michigan Agricultural Experiment Station, East Lansing, Mich.)

IMPROVING THE WILD BLUEBERRY, by Frederick

V. Coville. (*1937 Year-book Separate No. 1585*, U. S. Department of Agriculture, Washington, D. C.)

THE INSECTS OF THE CULTIVATED BLUEBERRY, by C S. Beckwith. (*Circular 311*, New Jersey Department of Agriculture, Trenton, N. J.)

PERIODICAL REPORTS ON BLUEBERRIES. (See "Cranberries," *The National Cranberry Magazine*, Wareham, Mass.)

*For blueberry culture in the home garden, see section on "Tillage and Mulch."

www.ingramcontent.com/pod-product-compliance
Lightning Source LLC
Chambersburg PA
CBHW032006220426
43664CB00005B/167